ALFRED ELY BEACH'S PNEUMATIC SUBWAY

THE SECRET TUNNEL BELOW NEW YORK CITY

By Kate O'Dell

FOR SASHA

A BRIEF INTRODUCTION

New York City has long been known for its underground subway system. New York's streets used to be filled with horse-drawn carriages, and were crowded, and congested. The story of how the city was able to eventually be transformed with its accessible public transportation system is a long, and complicated story. The history of those tunnels was filled with secretly built tunnels, corrupt politicians, innovative inventors, and a battle between various transit companies to push their plans forward. Eventually, the city's underground subway tunnels were built underneath the city, and the above-ground transit was installed.

Alfred Ely Beach - inventor, publisher, and patent lawyer - proposed a pneumatic subway system in the mid-1860s, basing it off London's Crystal Palace's experimental subway demonstration a few years prior. He was an inventor that was heavily invested in creating the city's first pneumatic subway tunnel. He secretly worked with a crew underground, building the tunnel, and the elaborately fancy subway station. When it finally opened in 1870, the subway was a hit with the public.

He worked diligently for years, trying to get the city's approval and funding he needed to expand his subway line. He wanted to make it five miles long, running all the way to Central Park. But ultimately, his efforts failed, when he was unable to get the backing.

Beach was a diligent worker. His inventions continued to be used for decades, and some are still used to this day. Alfred Ely Beach had a huge impact on the city's early transportation system, and his work was heavily documented in newspapers and scientific magazines throughout the man's career.

THE HISTORY OF THE BEACH PNEUMATIC SUBWAY

During the 1800s, New York City was known to be a city of contrasts. It was called that, because parts of it was huge, gorgeous brownstone buildings, while other areas had houses made of whatever material was on hand – often scrap metal, and wood. Some streets were made of cobblestones, while others were simply made of dirt. People used horse-drawn carriages, or walked on foot, as cars were not invented yet. That meant that the streets were often filthy – piled with horse dung, manure, and mud.

In the past, streets in New York were covered with stone-block paving, with flagstone crosswalks. They often had wide gaps between the stone blocks, and that unfortunately meant that manure would get caught in between them, and the smell would last for a long time. Though the stones were considered aesthetically attractive, it was difficult for horses to keep their footing on the smooth, shiny stones. They had poor traction, especially in certain weather conditions, like when it rained, or snowed, or when the streets were muddy.

Some areas of New York City still has small sections of visible flagstone crosswalks. One such area, is the intersection of Broadway, and Conway Street, located in Brooklyn. Another section of the city where the stones are still intact, is on the intersection of Ninth Avenue, and Little West 12th Street. The stones are lined up, some facing different directions in sections, so that it has a quilt-like appearance. Such is the case on Greenwich, and Gansevoort Streets.

The first known paved street in Lower Manhattan, was Stone Street. There are records of it being paved dating back to the 17th century. It was laid with cobbles – small stones that had been rounded by water. By the 19th century, the streets were a mess, because they were made with all sorts of varying materials. Some had Macadam, or gravel – but these were hard to clean, and created a large amount of dust. They also tended to break down under heavy use. Granite blocks were another material, preferred because they were easier to clean. Paving the streets with these required large teams of workers to get the job done. Often, politicians would pave streets with them close to election times.

During the 1850s, some streets had horsecar lines, so there were stone-block strips between the rails so that the horses could travel on them. The rails interfered with traffic, and there was a great deal of pressure to just pave the whole street with blocks.

In some areas, crosswalks had been installed, with large flagstone paths laid out at intersections. These allowed the pedestrians to cross the street easier, so that they wouldn't find themselves ankle-deep in mud, or manure.

Asphalt was introduced after the Civil War, introduced from Trinidad. It was considered the dream material, as it was smooth, and easy to clean. The horses were quieter on asphalt, than on the stone blocks. Health officials spoke out with their approval for asphalt, as it would make it easier to keep clean. The constant manure dust in the city left many with lung infections and diseases. Horses can produce about 15-35 pounds of manure per day, and so the streets were clogged with manure. And it attracted flies, which led to disease outbreaks, such as typhoid fever. Horse urine was also a problem, and also the horses had a life expectancy of 3-4 years. This meant that horses would sometimes die, and be left on the streets. In New York, there was a horse population of about 100,000, and that meant there was about 2.5 million pounds of manure per day.

The phrase 'traffic jam' had yet to be coined. However, the idea of traffic jam had easily been understood by New Yorkers. Traffic was at a standstill, impossible to get through at times. Broadway, in particular, was unmanageable. There were wagons, carriages, pedestrians, streetcars, and horse-drawn omnibuses. According to the New York Time, pedestrians faced daily gridlock that was so difficult to get through, that they found it nearly impossible to cross Broadway at any section below Eighth street.

It was one of the challenges of both pedestrian and driver alike. The streets were difficult to manoeuvre, mucky, and overcrowded. Something had to be done.

New York was always looking for ways to help with the city's traffic problems. One man, Charles T. Harvey, who was a self-taught engineer and inventor from Connecticut, had begun working on a shipping canal between Lake Superior, and Lake Huron. He was also working on railroad construction. He was the inventor of the elevated railroad. Harvey, and other inventors were trying to solve the city's traffic issues. The city wouldn't get its first proper underground subway, until 1904.

There were all sorts of ideas put forward to fix the city's transit issues. The *Arcade Railway* had been a proposed plan, that was vetoed by the Governor of New York. It was to be a high platform built above Broadway, that would have allowed for pedestrians, and horse-drawn vehicles to safely travel above the city street. There would have been large staircases on either side of the platform, and the streetcars would have been able to travel back and forth beneath the platform. Though it was an interesting solution, and one that could've potentially aided the city's traffic issues, it was never built.

Another idea, was '*Speer's Endless Traveling, or Railway Sidewalk*'. It was patented in 1871, by inventor Alfred Speer. It was a long, curved horseshoe-shaped road, that would've allowed for travelers to ride in specially-built chairs high above the traffic. It would've allowed for traffic to continue unimpeded below, while pedestrians could ride the slow-moving cars that were driven by a continuous screw mechanism along the track. The cars would travel through the stations dotted along the track, and would allow passengers to get on safely. Then, it would speed up between stations. It was a variable-speed system of rapid transit. Speers' plan was vetoed by the Governor.

In the 1850s, engineer James Swett had proposed an early vision for an elevated railway. His idea was to hang passenger cars directly underneath steam locomotive trains. It never got completed. The plan was not as well received as the others, and there were fears about the trains being directly above the passengers in close proximity.

The thing was, New Yorkers knew that they wanted a rapid transit system. It was more about what it would look like, and *how* it would be run, than it was about actually deciding on whether having a rapid transit system was a good idea. That was why there were so many differing ideas and plans to build the transit systems over the years. All sorts of engineers and inventors were vying to get their plans pushed through. Unfortunately, the moment that anybody's plans were actually close to being definite plans, they were forcefully opposed, and they were unable to complete their work. That is why there were so many scrapped plans from the 1840s onwards, and why most plans were turned down for public funding. Some ideas were simply turned down, because they were impractical.

Charles T. Harvey was able to design and build the world's first elevated street railroad on Manhattan's West Side. His elevated railway had train cars that were pulled by cables. London's first elevated railway had been built in 1836, but its construction included a brick viaduct, instead of being above an already existing and fully functioning railway.

Harvey's elevated track was built in 1867, over Greenwich Street. The traffic on rails, was supported by the thirty-foot-tall iron structure, high above the street and sidewalk. A year later, a cable-driven test line ran for half a mile on Greenwich Street, though it didn't allow for passengers.

The first ever elevated street railroad was finally opened to passengers. It only had two stops – Dey and Greenwich Streets, and at the Hudson River Railroad Terminal on 29th Street and Ninth Avenue. It was new, and exciting, and Harvey wanted it to succeed so much.

Unfortunately, the new system was plagued with problems. The cables snapped often, and that meant that passengers would be stranded. They would have to be lowered to the street by ladder. Within a few months, Harvey would be forced out of the company that he had founded. The cable cars were replaced with steam locomotives.

Harvey's idea had been an interesting one, but it hadn't worked in the long run. By the mid-1870s, all sorts of private companies were building steam-powered elevated railroads all over New York, and Brooklyn. And by the 1890s, the Manhattan Elevated Railway Company carried about 197 million passengers per year, and Brooklyn's lines carried about 30 million passengers annually. The *'el'* trains (elevated trains) became a part of New York life. The train trestles and tracks crisscrossing all over the city became defining architectural familiar landmarks around New York. Harvey was bitter that he'd been forced out, and other people had capitalized on his inventive idea.

The tracks left parts of the city shrouded in shadow, and there were other issues as well. The noise of the trains was constant and cumbersome, and the locomotives left smoke and ash in their wake – which was terrible for the environment, and made the air more difficult to breathe. There were also privacy issues, as people on the second or third floor apartments often complained that passengers would look into their units while they rode the trains. This was more often the case with lower class, or working-class tenants, as they didn't have a choice. In richer neighbourhoods, citizens fought to keep the elevated trains out. They fought by using political influence, as well as with lawsuits.

Amidst all this chaos, a man named Alfred Ely Beach believed that he had the solution to New York City's traffic congestion issues. Beach designed a functioning pneumatic subway system. It was the late 1860s, and the city streets were just as clogged as ever – full of wagons and horse-drawn carriages, merchants, and pedestrians struggling to not get run over on the crowded streets. Sometimes, traffic was so bad that drivers could only move forward a few yards in hours. It was chaotic. Drivers would sometimes even get into fistfights.

There was a lot of things to consider when designing his transit system. He had to decide if he wanted an above-ground system, or to go underground. The above-ground systems were loud, and people often complained that they didn't want the view obstructed. Besides, there was the soot and air pollution from the locomotive engines. He had been inspired by the Crystal Palace's exhibition a few years before, where they had demonstrated their underground Metropolitan Railway. Beach proposed creating the underground railway underneath Broadway, in hopes that it would help relieve the traffic congestion. Instead of using steam engines, which was common at the time, Beach wanted to try using trains propelled by a pneumatic system.

Alfred Ely Beach was a dedicated worker, and had already invented a great deal of interesting things before this. When he got an idea in his head, such as the idea of building the pneumatic subway, it was difficult to change his mind. This was what he was going to do, no matter what.

And so, Beach quickly got to work on his latest project. He was extremely excited at the prospect of using pneumatics. After all, he wouldn't need much – just a tube, a car, and a revolving fan to propel the passengers from one location to another. The carriages would be propelled using air pressure, which was generated by huge fans. It would be so much better than the soot-producing trains above-ground. He was certain of it.

By 1867, he had created a model of the basic pneumatic subway system, which used air pressure to push the cars through the tube. It went on display at the *American Institute Exhibition* in New York. Beach's prototype was an above-ground model to show people just how well it would work. He had built a large wooden tube that was about six feet in diameter, and 100 feet long. It could hold a small ten-passenger vehicle. His prototype had a huge ten-foot helix fan on one end, that was capable of blasting the air to propel the car through his tunnel. It made two hundred revolutions per minute. The fans would be reversed, and the car would be sucked back to the beginning of the tunnel again.

Alfred Ely Beach was able to demonstrate successfully that it was a viable model. Because of this, he established the *Beach Pneumatic Transit Company* in 1869. The public was definitely interested in his project, and Beach considered it a huge success.

The next step, would be to build a pneumatically powered subway line underneath Broadway. It was the very first attempt at building an underground public transit system in New York City. Beach, like all the other transit inventors, was met with pushback. It was one thing to have a prototype of the proposed system, but to build a tunnel and subway under the city? It would be a hard sell.

One of New York's top politicians at the time, a name named William "Boss" Tweed, refused to support the project. He hated the idea so much, that Beach had to start his project by claiming he was building postal tubes. That was what the initial permits said – that he was building a pair of smaller postal tubes underneath Broadway. His excuse, of building a pneumatic mail delivery system, was not completely implausible. Beach had, in fact, tinkered with said mail system – one that is still in use to this day in banks, stores, and some hospitals. It was the perfect sort of cover story.

Beach needed to find somewhere to work. Somewhere where he could come and go unimpeded, where he could keep the secret from the public until the subway tunnel was fully dug out. The exact location of the tubes was decided during construction, by using compass and survey. The location was also verified by driving jointed rods of iron up through the roof of the tunnel to the pavement.

Beach rented out the basement underneath the Rogers Peet building. It was an eight-storey building in the Civic Center, and Tribeca neighbourhoods in Manhattan. The building was home to the Rogers Peet clothing store from 1863-1898, until it burnt down. It later was rebuilt. It was located at Broadway and Warren Street, and belonged to the Emanuel Hoffman Estate. It was also directly across the street from City Hall, and they were none the wiser.

For the next few years, Beach worked in secret on construction of his secret project. He put up $350,000 of his own money to help build the line. Beach set about hiring crew that he could trust, people that would keep mum about his secret project. One of the people he hired, was his own twenty-one-year-old son, Frederick Beach.

There were all sorts of logistic issues when it came to building a subway tunnel in secret. The location was always teeming with people, and it made it incredibly difficult to get the work done without giving it away. They needed to bring in supplies, and make sure that nobody entered the premises who wasn't supposed to be there. Members of the public were curious, and there were often newspaper reporters interested in the underground tunnel.

The ground underneath Broadway where Beach had decided to dig out the tunnel was extremely sandy and loose. Beach considered all possibilities when it came to digging out the tunnel – especially when it came to the safety of his crewmen. The last thing he wanted, was for someone to get injured or killed in a cave-in, especially if the street above them were to collapse.

Beach invented something that he called a '*tunneling shield*'. A tunneling shield is a protective structure that is used when workers are excavating a large man-made tunnel. It is extremely important for the safety of the workers, especially for ground that is soft, unstable, or liquid. It helps protect them from falling materials, and from cave-ins. It is often used as a temporary support structure that is only in place until the tunnel can be lined with a permanent support structure. The sandy ground was the perfect sort of material for his hydraulic tunneling shield.

In 1868, the first load of dirt was excavated, and it was brought to the corner of Devlin's clothing store. Beach's son, Frederick Beach, was the gang foreman. Fred and his crew worked consistently on the underground wall of dirt. Many of the hired workers were scared of the working conditions under the clothing store. The job of digging out the tunnel was claustrophobic, and potentially dangerous. The workers could hear the horses galloping ahead, their hoofbeats loud in the tunnel. They feared that there would be a cave-in, and that they would be buried. Or that if there was a cave-in, the public would discover what they were really building. It was close quarters, and the workers dug out the sand and dirt in the dim lantern light.

At first, the workers were only digging through sand and dirt, and that was alright. They used their iron picks, cutting steadily through the dirt. But then one day, one of the workers hit a piece of stone. He tried to continue, thinking it was only a small chunk, but as it turned out, it was actually a wall of stone. They were forced to stop.

Beach and the workers believed that the stones were the foundation of an old fort. They thought that if they kept going, and removed the foundation, then the street above them might collapse onto them. Beach had to make a decision. Abandon his project, or continue. He decided to continue digging out the tunnel.

Slowly, ever so slowly, the crew managed to break through the stones. Beach was still incredibly worried that the ceiling would fall through, and he kept an eye on the ceiling. There was no sign of sagging, and Beach was relieved. He thought that he had made the right decision. They continued digging.

Every night, the workers would haul out the heavy bags of dirt and sand out of the tunnel. They had wagons specially made for the job, the wheels of the wagons were muffled to keep them silent. Once the wagons were wheeled away, crew would arrive with tools and bricks for the tunnel walls. The workers would continue coming and going from the tunnel, staying as quiet as they could. They needed to stay as quiet as possible, so as not to give the secret away. It was an extensive operation.

During construction, Beach's money had been dwindling. In the beginning of the project, he had put in his own money – to the sum of

$350,000. He needed to complete the subway by February 1870, and he was struggling to get everything assembled by then. Luckily, Beach managed to complete his tunnel in time. On February 26, 1870, he revealed his completed subway tunnel to the public.

For fifty-eight nights, the crew worked without incident. The bricks had been laid, the tunnel was dug out, and the tunnel was ready. All that he had to do, was finish putting in all the fancy fixtures. And soon, all of that was complete. New York was oblivious to the new subway tunnel, as was Boss Tweed. Only Beach, his son, and the crew he had hired knew what was really underneath the New York city street.

Throughout the next few years, while the tunnel was being constructed, newspapers wrote about Beach's tunnel. They wanted to know more, wanted to learn what Beach was up to. Beach was careful to keep the press at bay. People noted that there were materials being delivered to Warren Street, and so the newspapers reported on his secret comings and goings. Still, he refused to divulge any details. The details were only publicly revealed in an article published by the New York Tribune on January 11, 1870. It was a few weeks before the opening, and the story may have been planted to garner attention to the upcoming event.

One of the reasons why the project cost him so much, was because Beach was going all out. He was putting in all sorts of lavish fixtures – the grand piano, the frescos, the goldfish pond – all of it was reminiscent of the Metropolitan Opera House. Rich, full of splendour. Beach was a salesman at heart, and he believed that it was one of the main ways to win over the public. The subway had to be efficient, of course. But it had to be beautiful, and elegant, as well.

The station was a work of art, and extremely ornate. The waiting room that Beach designed, was 120 feet long. Instead of entering a dark, underground tunnel that was cave-like, Beach wanted the public to enter the elegant waiting room that was brightly lit. It was decorated with easy chairs, frescoes, and the subway car had zircon lamps to illuminate the station. Zirconia lamps are incredibly bright chemical lights that use zirconia. They aren't used anymore; the zirconia lamps have long since been replaced by electric light. There were all sorts of fun things for passengers to study while they waited to ride the subway. Aside from the piano, there were statues, and a goldfish pond.

The ride was controlled by a 48-short-iron Roots Blower. It was nicknamed the "Western Tornado". A Roots blower had two-lobed rotors, though many of them have three or four lobes. It was built by Roots Patent Force Rotary Blowers, from the Roots Blower Company. When the subway car reached the end of the tunnel, the baffles (baffles are flow-directing vanes or panels) on the blower system would be reversed, and the car would be pulled back by the suction.

Beach used a circular design for the tunnels. He based it off of Brunel's rectangular shield, and it was the shift in design from rectangular to cylindrical. For a while, it was unclear who shifted the tunneling shield design from rectangular to circular, until the New York Times ran an article that described the original Beach tunneling shield in 1870.

He was forty-four years old when the pneumatic subway was finally unveiled to the public. Beach was extremely excited about the subway's opening. The story had run in the newspaper, and the public were buzzing, talking about the new and exciting pneumatic subway tunnel.

The pneumatic transit system was opened to the public, and they charged a $0.25 fare per person. The admission proceeds went to the Union Home and School for Soldiers' and Sailors' Orphans. The subway car was able to hold 22 people at one time. The passengers would enter at Devlin's Clothing Store, which was a store located at 260 Broadway, on the southwest corner of Warren Street.

The plan for the subway system, was that it would run a total amount of 5 miles (8 KM), to Central Park. It would really help in reducing the heavy traffic above ground. Passengers could travel quicker, safer, more efficiently.

The public considered the subway project an attraction. It wasn't used as a proper form of transportation at that time. The subway only had a single car that ran on its one-block track to the dead-end terminus. Passengers would ride back and forth on the subway, trying to determine what the proposed subway would be like.

When compared to subway lines of modern times, Beach's subway was rudimentary, and it was not a regular mode of transportation. People still travelled by wagons, and carriages, and until vehicles became the main form of transportation, people still relied on horses to get around. That, and the steam locomotives that had begun cropping up in large cities (such as New York City, Philadelphia, Paris, London, etc.) during the 1800s.

In the first two weeks of operation, the Beach Pneumatic Transit sold more than 11,000 rides. In a single year, the subway held more than 400,000 rides. The public were enthused by Beach's idea, and enjoyed riding the subway. It was fun. A new, modern way of travelling. Things were looking up.

Beach invited all sorts of dignitaries, and the press, believing in the gamble that the grandeur of the subway would be enough to impress everyone. People were dazzled, alright. He was ecstatic at the thought that people were finally able to see all his hard work. They could ride the subway, and take in all the splendour for themselves.

In the press, there were all sorts of commendations about Beach's work. They focused on its grandeur, on the beauty of the subway tunnel. And in the *Scientific American*, they wrote about how it was nice to ride the subway without getting all the dust and grime in their hair, their clothes, their beards. It was a real concern, that the disgusting city smog would settle on them 'like a vapour on a frosty morning'. But with the underground subway system, that was not a concern. They could ride without dealing with the air's poor quality. In that time, it was known that pedestrians and passengers were barely able to make it ten blocks in downtown New York City, without suffering dizziness, or headaches from the poor air quality. This underground tunnel could be just exactly what the city needed.

The subway car was comfortable, with rich upholstery, and was incredibly spacious. Beach made sure to tell everyone that this level of decadence would be maintained for the future subway expansion. He told everyone who would listen, that the subway would be extended all the way to Central Park. That was five miles. He would be able to carry 20,000 passengers a day at speeds of up to a mile a minute. It was a single tunnel, and was 300 feet long (90 metres). It was eight feet in diameter, and it ran under Broadway, from Warren Street to Murray Street. The terminus at Murray street was not a station, but actually a dead-end.

By the time Beach's subway had been unveiled, there had already been an elevated railway built at Greenwich Street, and Ninth Avenue being constructed, that operated a small steam engine. It came into use in April, 1870. The elevated railway helped ease congestion. It gave James Henry Greathead the idea for the Docker's Umbrella in Liverpool, as they also built an overhead railway in England. Elevated trains like this were cropping up all over the city.

The waiting room at Beach's pneumatic subway garnered all sorts of positive attention. It was described in the press as being a large and elegantly furnished apartment – cheerful and attractive throughout. In fact, most everyone had something to say about the subway. That is, except for William Tweed. He stayed silent on the matter. His silence was troubling.

Word on the street, was that Tweed was equal parts stunned, and furious over the subway. He was unimpressed by the sudden unveiling of the tunnel – that had been built right under his feet this entire time. And he hated the fact that the public seemed excessively favourable towards Beach's project.

Beach's subway would cut into Tweed's profits. That was a problem. But there was also the issue of Beach deliberately going behind his back, and doing the one thing that Tweed had wanted Beach not to do. In

New York, every trolley car company paid tribute to Tweed. And now with this new subway, well, that meant that Tweed would lose out. He couldn't have that. Beach would pay for his secrecy, and for going up against Tweed.

Tweed was a powerful political figure in the city, with a lot of connections. He was a powerful foe. Tweed needed to find a way to bring Beach down, and fast. And as soon as Beach heard that Tweed was going after him, he warned Tweed that he was willing to go before legislation in Albany. Anything to keep his subway tunnel running. There was no way he was going to allow all his invested money, and hard-earned money to go to waste.

Beach had the public's approval to his advantage. Everyday, people would line up in hordes to try out the new and exciting subway. They would gawk at the beautiful waiting room, the frescoes, the grand piano. For months, they would pay the quarter fare to ride the subway. And when the subway was not running, they would have the opportunity to walk through the newly-constructed tunnel, and admire its handiwork.

Beach went to the legislature. The Beach Transit bill had a hefty five-million-dollar price tag. All of the money had to be privately raised. And the work underground had to be done with as little disruption to street level as possible. It was a win for Alfred Ely Beach. The New York State Senate passed the bill in a landslide 22-5 vote. The State Assembly gave it their stamp of approval. There was only one thing. Tweed was still bitter about the subway, and determined to viciously bring him down.

Tweed had an idea that would go against Beach's subway. He had called it 'Tweed's Viaduct Plan', and the blueprints showed elevated lines that were mounted on forty-foot-high stone arches. The project would be publicly funded, and would cost eighty million dollars.

The Viaduct Plan was approved in the state legislature. The Governor – John T. Hoffman – would need to decide which plan would go through. Beach's, or Tweed's? As soon as the plans went across the Governor's desk, Beach's plans were doomed. He vetoed the plans, and pushed Tweed's through instead. The press wrote articles and editorials about how Beach had been treated unfairly, and he still had public support. But it was too late.

Beach knew that he needed to get support from the public, as he campaigned for his subway. He wanted the Governor's veto to be overcome. It was a difficult battle, as Tweed had a lot of politicians in his pocket. The man was a notoriously dirty fighter. The only official that ended up accepting Beach's invitation to ride the subway, was the Secretary of the Navy Robeson. He rode the subway on December 1870. He really liked the subway, and spoke to the press about how it was a great subway. His word went a long way. Still, in Beach's eyes, it wasn't enough publicity.

The public still enjoyed Beach's subway, and when the Beach Transit bill came up for another vote, it passed. And then with a stroke of luck, it received Hoffman's expected veto. The legislature tried to override the veto. It needed a two-thirds majority. The votes were tallied, and Beach lost by one vote. It was a terrible blow. Beach could've given up then, but he was an extremely determined man. He kept fighting. The subway meant that much to him. He wasn't going to let a corrupt man like Tweed win.

In 1872, The New York Times started to stories about corruption at Tammamy Hall, which directly affected Tweed and his empire. Tweed was indicted. (He went to jail, later escaped to Spain, and then was returned to the United States to finish his sentence. Tweed died of pneumonia in prison, on April 12, 1878). In November 1872, Governor Hoffman was voted out of office.

Though people still loved the subway, the pneumatic tube had begun to lose its appeal. It had been two years, after all, and the subway was hardly making the amount of money through admissions as they once had been. Beach had to close the subway for a short while, because it wasn't as feasible as it had been during its opening. He was still fighting tooth and nail to keep it running, to get all the funding that he needed.

He met other obstacles. Engineers had begun arguing with Beach about how his hydraulic tunneling shield wouldn't be as effective through Manhattan's rocky ground. There were other scientists that believed Beach's subway would be ineffective on longer runs, such as the five-mile trek that Beach had planned. The pneumatic power seemed to be only enough for the short subway. And so, Beach had to re-examine his subway plans. If the pneumatic tube didn't work, he decided to allow steam engines to pull the trains. And if his shield didn't work? He had plans for that, too. He'd switch to a different method – cur-and-fill technique.

Local landlords, such as millionaire John Jacob Astor. Astor and other landlords feared that the subway tunneling underneath Broadway could endanger building foundations. Trinity Church, with its 280-foot tower, was Manhattan's tallest building at the time. That one might be at risk, as well as other buildings in the area. They opposed Beach's plans, and fought against his work.

Beach had issues with finding sufficient funds for his project. He was struggling to find the money, because inflation meant that instead of seeking five million dollars, he needed to get *ten* million. The money needed to be raised by such people like Astor. Astor didn't want any part of his subway. Beach went to all sorts of financiers, asking for the funds. One by one, he was turned down. However, Beach needed to get permission to expand the subway line. He had a hard time getting permission for a variety of reasons. He finally managed to get permission in 1873, but by then, financial support had lessened – as had public support. This meant that unfortunately, the subway shut down within a year. There was a stock market crash, which caused investors to pull their support. Some believed that the system might've worked as a large-scale subway network, but others were not so sure.

In late 1873, Governor Dix officially withdrew the charter for Beach's pneumatic subway. And so, Beach's subway never got completed.

His project had failed. Beach was devastated. The subway line only lasted from 1870-1873.

When the panic of 1873 hit, Beach was finally forced to call it quits on his pneumatic subway system. People who had invested in the stock market had lost everything. People were left destitute. And some of those who suffered such a financial loss, were the people backing his transit system. The failed pneumatic subway system had been years in the making, as he had had to fight to get investor's support, as well as struggling because certain businessmen were opposed to the subway's construction. They feared that it would damage buildings above ground.

Though the rich businessmen had been opposed to Beach's underground subway system in Broadway, they were not opposed to the railway built away from Broadway. By the mid-1870s, elevated railways were being considered practical, opposed to the underground system that Beach had proposed. During the 19th century, New York City continued to develop rapid transit elevated railways all over the city.

Unfortunately, the British pneumatic tube system failed to be as popular as they had hoped. Royal Mail had been contracted to use the tunnels, but they soon fell into disrepair. It was considered a failed English experiment, because of lack of proper funding, and also because of technical issues.

When the project got shut down, the tunnel entrance was sealed. The station that was built in the basement of the Rogers Peet Building, was used for other uses until the building burned down in 1898. For example, the basement was used as a shooting range at one point, and the basement was rented out to other tenants.

Elevated trains became the norm around New York City. The city of Brooklyn became amalgamated into New York in 1898, as well as other areas – and it was considered a borough. Bridges were built, people were starting to buy cars. The city's landscape went through a variety of changes as time went on.

In 1912, workers were excavating for the BMT Broadway Line (which now serves the present-day N, R, and W trains), dug into the old Beach Tunnel completely by accident. They had had no idea that it was there.

They were shocked to discover the rich and glamourous subway tunnel buried underground. Some of the waiting room's wooden fixtures had rotted. But for the most part, it was just as appealing as it had been at its opening. The workers found the old subway car, and the tunnelling shield used when they excavated the tunnel. They even found the piano that was used in the subway's waiting room. The tunneling shield was removed, and it was donated to Cornell University. Since then, the University has lost track of the shield's whereabouts.

The tunnel was nearly completely within the limits of the Broadway Line's City Hall station. It is rumoured, however, that there is a small section of the subway line that can still be accessed by the manhole on Reade Street.

The New York Historical Society commissioned a plaque to honour Alfred Ely Beach. It was placed in the City Hall station. Although the Beach Pneumatic Transit had only lasted for three years, the project had a lasting impact on the way the city ran things. The New York pneumatic tube mail system was based on Beach's work. It continued to run in the city until 1953.

PHOTO GALLERY

Here is a photograph of Charles T. Harvey riding his elevated railroad above Greenwich Street, in 1867.

This is a photograph of the Beach Pneumatic Transit, in 1870.

Here is an example of a tunnelling shield. This is the side view of the shield that was used to build the Thames tunnel. The permanent brick supporting structure is build immediately afterwards. This illustration is from the 19th century, and was published in the Illustrated London News.

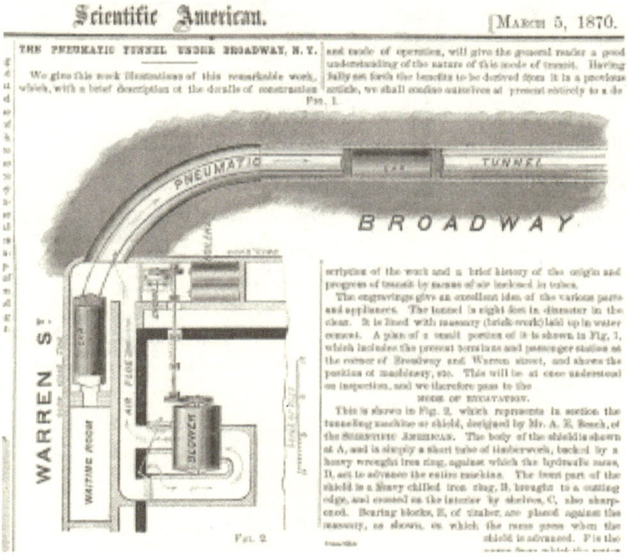

Scientific American. [March 5, 1870.

THE PNEUMATIC TUNNEL UNDER BROADWAY, N. Y.

We give this week illustrations of this remarkable work, which, with a brief description of the details of construction and mode of operation, will give the general reader a good understanding of the nature of this mode of transit. Having fully set forth the benefits to be derived from it in a previous article, we shall confine ourselves at present entirely to a description of the work and a brief history of the origin and progress of transit by means of air inclosed in tubes.

Fig. 1.

The engravings give an excellent idea of the various parts and appliances. The tunnel is eight feet in diameter in the clear. It is lined with masonry (brick-work) laid up in water cement. A plan of a small portion of it is shown in Fig. 1, which includes the present terminus and passenger station at the corner of Broadway and Warren street, and shows the position of machinery, etc. This will be at once understood on inspection, and we therefore pass to the

MODE OF EXCAVATION.

This is shown in Fig. 2, which represents in section the tunneling machine or shield, designed by Mr. A. E. Beach, of the SCIENTIFIC AMERICAN. The body of the shield is shown at A, and is simply a short tube of timberwork, backed by a heavy wrought iron ring, against which the hydraulic rams, D, act to advance the entire machine. The front part of the shield is a heavy chilled iron ring, B, brought to a cutting edge, and crossed on the interior by shelves, C, also sharpened. Bearing blocks, E, of timber, are placed against the masonry, as shown, on which the rams press when the shield is advanced. F is the

Fig. 2

This is the plan of the Beach Pneumatic Transit Station and Tunnel. It is a scan of the original copy, that appeared in the Scientific American *(March 5, 1870).*

This was the only surviving photograph of the Beach Pneumatic Transit, while it was still in operation. It was taken in early March, 1870, by Rockwood & Co. The photograph was one of a series of photos taken of the pneumatic system. They needed to use artificial light to photograph it, as it was entirely underground. They used large oxy-hydrogen calcium lights. They included the hanging lamp above, to give it a more 3D (stereoscopic) effect. The blurry parts around the top of the arch were the coloured gas jets. The subway car was just inside the whitewashed tunnel.

This illustration appeared in Frank Leslie's Illustrated Newspaper, *on February 19, 1870, and in* Scientific American, *March 5, 1870. It showed the comfortable seating in the subway car, the two zircon lights, and the two running rails, and the center brake rail.*

The sketch above appeared in Harper's Weekly, on March 12, 1870. It also appeared in Beach's Illustrated Description. It shows an above view from the waiting room down to the subway car. This was on the Broadway side, between the building and the vault. They had the window instead of an opening, to maintain pressure in the station. The car needed to start its turn in order to enter the tunnel.

This illustration appeared in Frank Leslie's Illustrated Newspaper, on February 19, 1870, and in Scientific American, March 5, 1870. It shows the tunnel entrance, before it had its lighting fully set up. The illustration shows the short brick section through the basement wall, as well as the Dixon patent iron plates along the curve, and the track construction. In later illustrations, the keystone contained the words 'Pneumatic / 1870 / Transit', and the tunnel floor was covered properly to allow people to walk inside more comfortably.

This view looking back to the station shows the car where it had stood, when the tunnel was open for walking. There was a cover for the brake rail. The lighting and fixtures were

installed, and the floor was properly covered. It appeared in Appleton's Journal of Literature, *on June 25, 1870.*

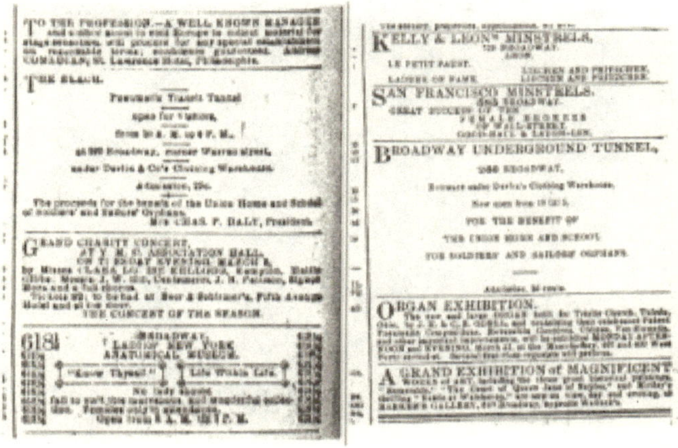

TO THE PROFESSION.—A WELL KNOWN MANAGER

GRAND CHARITY CONCERT,

THE CONCERT OF THE SEASON.

KELLY & LEON'S MINSTRELS,

SAN FRANCISCO MINSTRELS.

BROADWAY UNDERGROUND TUNNEL,

FOR THE BENEFIT OF

ORGAN EXHIBITION.

A GRAND EXHIBITION of MAGNIFICENT

The image on the left, was the first ad for Beach's subway tunnel. It appeared in the Herald, *on March 5th, 1870. On the right, was the ad from the* Tribune, *on March 21st, 1870. They were both posted in the 'amusements' section.*

This was the view down the long tunnel, showing the brake rail that was covered around the curve when the subway car wasn't running. This was the end of the iron plate curve, going down Broadway to the end of the tunnel. The man's stance depicts the uneven floor level. It appeared in Harper's Weekly, *on March 12, 1870.*

The above illustration appeared in Appleton's Journal. *It is an image depicting the waiting room. There were paintings on the wall, and the grand piano and fountain are just out of sight. There was daylight coming into the tunnel via round openings in the ceiling, which were under the sidewalks on Warren Street. The opening on the left is where the blower was.*

This was the prototype that Alfred Ely Beach had used for his demonstration, in 1867. It was on display at the American Institute exhibition, on Fourteenth Street.

ALFRED ELY BEACH

ALFRED ELY BEACH WAS born on September 1, 1826, in Springfield, Massachusetts. He was the son of Moses Yale Beach, who was an American inventor and publisher. Moses Beach started the Associated Press, and credited with starting the print syndication in 1841.

He went to Monson Academy, in Massachusetts. The academy has since been renamed, and is now 'Wilbraham & Monson Academy'. Inventing was in his blood. Beach was an inquisitive person, and always looking for innovative solutions to everyday problems. He was known to be a hard worker. In later years, he was known to be such a hard worker, than he never took vacation time. Beach was highly intelligent, and a well-read man.

Alfred Ely Beach worked for his father, until he and his friend, Orson Desaix Munn, decided to buy a new publication titled '*Scientific American*'. The pair ran the publication until their deaths years later, and then it continued to be run by their sons, and grandsons for years.

Aside from being an American inventor, he was also a publisher, and Munn and Beach had also created a successful patent agency. They had created all sorts of interesting inventions, such as an early typewriter for the blind. His typewriter invention won him a gold medal at the Crystal Palace Exposition, in 1853. Beach and Munn created one of the world's first tunneling shields the same year as famed engineer James Henry Greathead.

When the civil war was over, Beach also founded a school for freed slaves in Savannah. It was called the Beach Institute. It shut down in 1919, when public education was finally made available. The building remained, however, and was used for various non-profits, and the Savannah Boys' Club. It is now know as the *Beach Institute African American Cultural Centre*, and is managed by the '*King-Tisdell Cottage Foundation*'.

This is an illustration of the Beach Institute, as it appeared in Harper's Weekly, on October 3, 1868.

BEACH WAS BEST KNOWN for the design of New York City's early subway system (the Beach Pneumatic Transit). He created the idea in the late 1860s, because traffic in New York at the time was horrible, and that was especially true in the Broadway area. The roads were incredibly loud, full of horses neighing, and wagon drivers shouting, and people struggling to get through the clogged streets.

Beach knew that there were only two options – have an above-ground system, or tunnel under the city streets. He didn't like the above-ground system, because he thought it would be too loud, and a blight on the landscape. Besides, there was always the issue of too much soot – which was terrible for the air pollution. It was difficult to determine how to build the subway. Gasoline and electric motors were not yet options. He decided that pneumatic power was his only option.

Beach proposed creating the underground railway underneath Broadway, in hopes that it would help relieve the traffic congestion. He had been inspired by the underground Metropolitan Railway in London. Instead of using steam engines, which was common at the time, Beach wanted to try using trains propelled by a pneumatic system. He used a tunneling shield during construction, so that it would minimize disturbing the street above while they worked.

Aside from his pneumatic train system, Beach also wanted to use pneumatic tubes to transport letters and packages – as this was something that he knew was put to use in London. He started using pneumatic tubes in 1869. His demonstration of his pneumatic subway opened on February 26, 1870.

For the next few years, Beach worked in secret on construction of his secret project. People noted that there were materials being delivered to Warren Street, and the newspapers reported on his secret comings and goings. Still, he refused to divulge any details. The details were only publicly revealed in an article published by the New York Tribune on January 11, 1870. It was a few weeks before the opening, and may have been planted to garner attention to the upcoming event.

When the panic of 1873 hit, Beach was finally forced to call it quits on his pneumatic subway system. It had been years in the making, as he had had to fight to get investor's support, as well as struggling because certain businessmen were opposed to the subway's construction. They feared that it would damage buildings above ground.

There was an elevated railway built at Greenwich Street, and Ninth Avenue, that operated a small steam engine. It started in 1870. The elevated railway helped ease congestion. It gave James Henry Greathead the idea for the Docker's Umbrella in Liverpool, as they also built an overhead railway in England.

Though the rich businessmen had been opposed to Beach's underground subway system in Broadway, they were not opposed to the railway built away from Broadway. By the mid-1870s, elevated railways were being considered practical, opposed to the underground system that Beach had proposed. During the 19th century, New York City continued to develop rapid transit elevated railways all over the city.

Unfortunately, the British pneumatic tube system failed to be as popular as they had hoped. Royal Mail had been contracted to use the tunnels, but they soon fell into disrepair. It was considered a failed English experiment, because of lack of proper funding, and also because of technical issues.

The Beach Tunneling Shield, was later used to build the Grand Trunk Railway of Canada's first St. Clair Tunnel (it ran from Port Huron, Michigan, to Sarnia, Ontario). The tunnel opened in 1890.

Beach decided to donate land to seven men who were trying to start a yacht club in Stratford, Connecticut. This took place in January, 1887. It is the oldest operating yacht club in Connecticut, on the Housantonic River.

Alfred Ely Beach married Harriet Eliza Holbrook on June 3o, 1847, in Boston, Massachusetts. She died on January 8, 1905, in Stratford, Connecticut. She was 78 years old. Harriet was buried at Union Cemetery.

Alfred Ely Beach and Harriet Eliza Holbrook had a son. He was named Frederick Converse Beach, and was born on March 27, 1848, in New York City. Frederick was editor of the *Scientific American* magazine, and the *Encyclopedia Americana* in the early 1900s. Like his father, and grandfather, Frederick was an inventor. He invented the photolithographic process. Frederick died on June 8, 1918, in Stratford, Connecticut.

Beach had a daughter named Jennie Beach, who was born on October 28, 1850, in Stratford, Connecticut. She was the daughter of Alfred Ely Beach, and Harriet Holbrook Breach. Jennie married Marquis Cossitt Gasper, and she became stepmother to his five children from a previous marriage), one of whom was actually a year older than her. She died on February 25, 1932, in Stratford. She was buried at Union Cemetery, in Stratford.

Alfred Ely Beach died of pneumonia on January 1, 1896, in New York City. He was 69 at the time of his death. He was buried at the Union Cemetery, in Stratford, Connecticut.

SIMILAR PNEUMATIC RAILWAY SYSTEMS

There was a similar pneumatic railway system known as the Crystal Palace pneumatic railway, in London. It operated from August-October 1864. It was an experimental atmospheric railway that ran for 600 yards (550 metres).

It was designed by Thomas Webster Rammell, who had built a pneumatic railway for the London Pneumatic Despatch Company (it was built to have letters travel along tunnels in large vacuum-driven wagons). Rammell used similar techniques when he built the railway, as he fitted a carriage along an airtight tunnel that measured ten by nine feet.

It was a huge red and terra cotta brick building, that had large glass windows, and an iron train shed roof. The building had square towers, each with short spires on each corner. There were brick arches inside, separating the interior of the station. There was a passenger concourse set above the tracks, that held waiting areas, a booking office, and refreshment rooms.

The railway system only ran for a little over two months, and it was more of a demonstration line, then a fulltime sort of transportation. There was a subway line planned between Waterloo and Whitehall, to be built underneath the Thames, but it was never completed. The tunnel was a shallow trench, only about four feet in depth.

The above illustration appeared in Alfred Ely Beach's 1868 book 'The Pneumatic Dispatch, with Illustrations'. It depicts the Pneumatic Dispatch, in Battersea Fields, London, in 1861.

THE TUNNEL RAN FOR 600 yards, between Sydenham and Penge entrances to the park. Tickets were sixpence each. It would run during the day (from 1 PM to 6 PM), and the entire journey only took 50 seconds. It operated from August-October 1864.

The tunnel was later destroyed, possibly to make way for the Festival of Empire celebrations in 1911. For years, people believed that the site of the railway was haunted. There was an urban legend in the 1930s that claimed the Crystal Palace railway station was haunted. It was believed that a commuter train had been trapped in the tunnels after there was a collapse. Workers were apparently unable to get the passengers out safely. When they realized they couldn't get them out safely, the tunnel was permanently bricked up, leaving the passengers to die. It is said that the ghosts of these trapped passengers still haunt the area to this day. It is unsubstantiated, though.

In 1978, a woman claimed to have found the missing tunnel. She said that she'd found a railway carriage filled with skeletons wearing old Victorian outfits. The legend took on a life of its own. Fantasy writer used that legend as inspiration for his novel, titled 'Strange Air'.

ANOTHER EXAMPLE OF an abandoned New York tunnel, is the Cobble Hill Tunnel. It ran beneath Atlantic Avenue in Brooklyn, New York, and ran through downtown Brooklyn, and Cobble Hill neighbourhoods. It ran for 2,517 feet (767 metres), between Columbia Street, and Boerum Place. It was the oldest railway tunnel underneath a North American city street, that was completely devoted to rail.

It was originally built as an open cut type of construction, starting in May 1844. It was opened to the public in December 3, 1844 (even though it wasn't actually completed until the middle of 1845). About five years in, the open cut was turned into a tunnel with a roof installed. At first, it had been 21 feet wide, and 2,517 feet long. Once it was roofed over, the tunnel's height was 17 feet tall.

Brooklyn allowed permission for steam locomotives on Atlantic Street west of Fifth Avenue (which at the time, was Parmentier's Garden/ Gowanus Lane), all the way to Brooklyn's South Ferry. The train cars were pulled by huge teams of horses along Atlantic Street, from Clinton Street to Parmentier's Garden. The steam locomotives were attached. It was part of a rail link between New York City, and Boston, Massachusetts.

The ends of the tunnel were sealed up in Autumn 1861. The reason why it was shut down, was because the New York State Legislature voted to ban all railroad locomotives within the city limits of Brooklyn.

(Brooklyn was a city, until 1898, when it was consolidated into New York City, and turned into a borough).

There was a tax assessment done along all property owners along Atlantic Street, to defray the costs of the closure. What was not commonly known at the time, was that New York State Governor John A. King was a huge shareholder in the Brooklyn and Jamaica Railroad (which was later called the Long Island Railroad). This meant that it was a conflict of interest, and that he stood to make money from the compensation payments to the railroad from the tax assessment. The Murray Hill Tunnel, which was built as an open cut around 1836 (and was roofed over in 1850s), is now in use for vehicles.

The tunnel stayed closed for years. That is, until 1916. The FBI believed that German terrorists were secretly making bombs in the tunnel. They decided to break through the roof of the tunnel with jackhammers, to catch them in the act. They didn't find any terrorists, or bombs. It was empty. They installed an electric light in the tunnel, then resealed it.

During the 1920s, it was rumoured that people were growing mushrooms in the tunnel, and that there were bootleg whiskey stills down there. The tunnel was still sealed, and the public didn't have any access to the tunnel. The tunnel had taken on a life of its own, being the object of all sorts of local legends and folklore.

The police believed that there was a hoodlum buried in the tunnel, and in 1936, they tried - and failed - to break into the tunnel to look for the body. The FBI broke in again (this time during the 1950s), because they were searching for spies. They didn't find any. For years, people had been looking for the tunnel, trying to see what secrets it held. It wasn't until 1980, that the tunnel was rediscovered.

A twenty-year-old man named Robert Diamond crawled into a manhole on Atlantic Avenue and Court Street. He crawled 70 feet (21 metres) underground through a section of tunnel that had been filled in. It was only two feet high at that point. He found the bulkhead wall that had sealed off the main part of the tunnel. He sought help from the Brooklyn Union Gas Co. engineering crew, and the were able to break through the huge concrete bulkhead wall. It was difficult work, as the wall was several feet thick.

Robert Diamond was able to open access to the main part of the tunnel. The tunnel has been listed on the National Register of Historic Places since 1989. He was able to start hosting public tours of the tunnel. He did so from 1982, to December 17, 2010. The Department of Transportation decided to terminate his contract, because they feared that it wasn't safe.

In 2013, Elon Musk proposed a hyperloop system, that he hoped would one day propel capsules through tubes at high speeds. He wanted to use magnetic levitation, and linear electric motors. This system has yet to be built.

POP CULTURE AND THE PNEUMATIC SYSTEM

People have been intrigued by Beach's pneumatic system since its development. The pneumatic railway system has appeared in all sorts of books, TV shows, and movies.

The Beach Pneumatic Transit was featured in the film 'An American Tail: The Treasure of Manhattan Island'. It is a large part of the plot. The 1998 film featured Fievel and Tony, as they discovered that an ancient treasure was underneath Manhattan. They went down to the abandoned Beach Pneumatic Transit system, and found a treasure map. This leads the tiny mice on a grand adventure, as they get assistance from an archaeologist, and search for the treasure.

There is a children's fiction series titled 'Pyrates' by Chris Archer. It is about a group of kids who find a pirate treasure map, and begin to travel in secret underground subway tunnels. The book features Alfred Ely Beach's pneumatic subway tunnel.

Another fiction book that features Alfred Ely Beach, and his subway tunnels, is the book 'Downsiders' by Neal Shusterman. It features a group of people who survive underground in a series of tunnels, that turned out to be the same tunnels that Beach had dug out more than a century before. In the book, Beach had faked his own death, and hidden underground. It is a very interesting book, and one that is great for young readers interested in Beach's tunnels.

There is a song by Klaatu, titled 'Sub-Rosa Subway', that describes the subway. The song describes Alfred Ely Beach's efforts on building the subway. It was called 'sub rosa', because the title references Beach's secretive work. The song hit the Billboard Hot 100 chart in 1977. In one portion of the song, there is a secret morse code message. It read as follows:

"From Alfred, heed thy sharpened ear – A message we do bring – Starship appears upon our sphere – Through London's sky come spring."

In the 1989 film 'Ghostbusters II', there was a fictionalized pneumatic transit station, and tunnel, that was similar to Beach's subway. The tunnel is discovered by the Ghostbusters underneath First Avenue in Manhattan. In the film, the tunnel's completion date was said to be 1870. This was the same year that Beach's tunnel was completed.

In the 1990 film, the Teenage Mutant Ninja Turtles used an abandoned subway tunnel system where the turtles lived. It was in reference to Beach's pneumatic tunnel.

Shana Corey's 'The Secret Subway' is well-written children's book about the pneumatic subway. It perfectly describes the history of New York during the 1860s-1870s, and Alfred Ely Beach's efforts to build the pneumatic subway system.

Alfred Ely Beach wrote a book in 1868, titled 'The Pneumatic Dispatch, with Illustrations'. It is an extremely good read for those that are interested in learning more about the pneumatic subway systems all over the world. It is especially of interest, because it was written by Alfred Ely Beach, and is complete with illustrations of the London trains, along with the ones in New York City. It has illustrations, and explanations of just how the proposed subway would work.